Dedicated:
to all those who gain pleasure,
appreciation or understanding
with every walk in the rainforest

CONTENTS

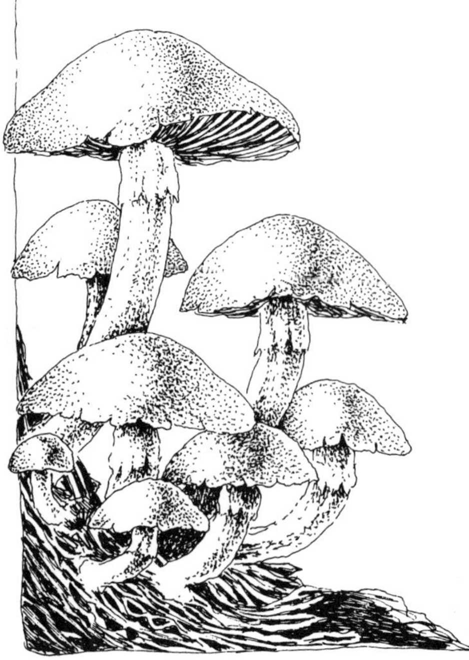

Forest floor fungi.

A WALK IN THE RAINFOREST

an introduction to the environment

by

Dawn and Clifford Frith

Illustrated by

David Stacey

Pale yellow robin.

Lake Barrine

Published and designed in Australia by Tropical Australia Graphics,
Paluma via Townsville, North Queensland 4816
Produced by Topcat Graphics, Townsville, Queensland 4810
National Library of Australia Card Number ISBN O 9589942 O X

First printed 1985
Second printing 1987

Macleay's honeyeater.

Green tree frog.

INTRODUCTION

The aim of this small book, delicately illustrated by David Stacey, is to indicate to the reader the beauty, fragility and intricacy of one of Australia's natural assets — a natural heritage that covers less than one quarter of one percent of her land mass and yet houses well over fifty percent of all her plant and animal life. Of the mere two million hectares of rainforest throughout the nation one million occurs within the State of Queensland.

Queensland's rainforest covers less than one eighth of one percent of the State and occurs in remnant patches from Lamington in the south to the Cape York Peninsula in the extreme north along narrow coastal plains, foothills, mountains ridges and tablelands. Only half of what was here prior to the arrival of white man now remains, and unfortunately is still continuing to disappear or be degraded. It is the lowland rainforest, richest of all, that has been most severely damaged or cleared completely over extensive areas for agricultural purposes and in particular sugar, a crop of which the world now has a glut. Only along the coast north of the Daintree River does the last extensive stand of coastal lowland rainforest remain relatively intact. It is also the only area where rainforest extends from mountain top, down over foothills, and right down to the coast to meet mangrove forests and fringing coral reefs. Magnificant stands of upland rainforest can still be seen around Lamington, Eungella, and Paluma Range to Atherton Tableland and north to Cooktown but it must be pointed out that very much of this has been severely logged in the past. It is the virgin rainforest which is so much richer, valuable and easier to traverse to us; but sadly all too little of which remains.

We have tried to describe plants and animals in this book in simple language, avoiding scientific words and names, and to make our comments about them applicable to all of Queensland's rainforests. Generally, rainforests south of the Tropic of Capricorn are referred to as subtropical and those north are of course the tropical ones. Those in the subtropics tend to be less diverse than those further north, although their basic structure is similar. Within each of the two basic categories are several different types of rainforest. For example lowland rainforests are characterized by large-leaved plants, robust woody lianes, lovely fan palms and strangler figs; whereas in the uplands tree ferns, epiphytic ferns, mosses, climbing vines as well as woody lianes are notably abundant. Some animals occur in all rainforested areas, but others, as noted in our text, have a more restricted to extremely restricted range. For example, the Southern Cassowary occurs only in wet tropical forests; the Regent Bowerbird is found in the rainforests of Eungella and southwards; and the Palm Cockatoo can only be seen on the Cape York Peninsula.

Ulysses (above) and Australian birdwing butterfly on lantana.

Travelling along a road or well cleared track through Queensland rainforest is how most people see it and, whilst this is a pleasurable experience, this does little to permit one to appreciate the incredibly rich and complex diversity of this the world's richest habitat or ecosystem.

The edges of rainforested roads are in fact colonized by secondary regrowth species of shrubs and trees such as the sarsaparilla tree (above), brambles, bracken and lantana thickets. Lantana (page 9) is an introduced species of South American origin that grows so quickly and efficiently that it has rapidly become a major pest in badly disturbed rainforest areas. It blankets the forest floor so quickly that it prevents the germination and subsequent regrowth of the naturally occurring rainforest species, which of course in turn support our native wildlife. Some rainforest-dwelling butterflies such as the Australian Birdwing and blue Ulysses Butterfly can often be seen sipping nectar from the colourful red, orange and yellow flowers of lantana (page 9), or Crimson Rosella parrots (opposite) may be spotted eating the green fruits. The presence of this flower nectar source also attracts and supports many butterflies not otherwise found in rainforest areas.

Stinging shrubs are native forest plants but, like lantana, favour more open sun-lit situations. In areas subjected to much human disturbance they thrive very well, often to the point of becoming a troublesome invader. Watch out for stinging shrubs when walking along a poorly cleared or ill-defined track for their leaves (below) and stems are armed with tiny stiff glass-like hairs that, if brushed against, will inflict a severe painful sting.

The structure of the rainforest itself is altered severely where roads and wide tracks are cut through it, when trees are felled, or even when damaged naturally by cyclones. In these disturbed areas rainforest becomes abnormally dense, often impenetrable, comprising of thickets of shrubs, small trees, vines and woody lianes entangled together with invader species.

Unfortunately, it is the disturbed aspect of rainforest that most people see, unless prepared to enter it and start delving beyond the disturbance, and it has sadly resulted in the common misconception that rainforest is a hostile, lifeless, inaccessible scrub; certainly not an environment to explore and enjoy let alone walk through for any length of time. By walking through just one Queensland rainforest (page 48) it is possible to see a good cross-section of its unique plant and animal life. Using this illustrated account as an introductory guide step into the rainforest with us and enjoy the beauty of one of Australia's magnificant wilderness areas. We explain very simply the fundamental ecology of rainforest life, how plants and animals are closely inter-related and how different species of animals live harmoniously together. It is hoped that this brief introductory account will provide a greater understanding of the complexity of plant and animal life in this most intricate of earth's ecosystems. Like all complex, finely developed and balanced things, the rainforest is extremely fragile. Contrary to the widely held view, rainforest is not a dangerous impenetrable green fortress but it is a most delicate and easily disturbed structure of vegetation standing upon extremely shallow and poor soils that supports much of unrealised value to mankind.

Most of all it is our hope that this brief informative sojourn into a rainforest will stimulate an interest in the conservation of this priceless, irreplacable, natural, national and international heritage, for both future generations of Australians and visitors from all over the world to enjoy.

Crimson Rosella.

THE FOREST FLOOR

When entering a rainforest one is immediately struck by a marked change in climate. Within the forest interior it is noticeably cooler, damper and darker, particularly near to the forest floor. The overshadowing, interlocking, leafy branches of the canopy provide and maintain this atmosphere in which plant and animal life flourishes in great abundance and diversity.

Old leaves flutter down from the canopy through the lower forest to the floor; rotten branches snap off and crash to the ground; and ancient or diseased trees topple over, smashing down other vegetation in their wake. Mosses in all shades of green, and fungi of all shapes sizes and colours soon blanket the fallen and decomposing plant material, until it blends into the forest floor.

Bacteria and fungi are the instigators of the decomposition process. What we see and call fungi are in fact the reproductive fruiting bodies of the plant. These bodies contain billions of single-celled reproductive spores that when mature are dispersed by wind, rain-splash, animals or by the structures themselves exploding. The Maiden Veil or Stinkhorn Fungus (below) emits the terrible smell of rotting meat to attract carrion flies which then function as dispersal agents for the fungal spores which are carried off by the misled flies. Although many of these fruit bodies resemble the classic mushroom shape some take on quite bizarre shapes resembling corals, staghorns, brackets, shells and tongues. Their, often quite vivid, red, orange, yellow, blue and purple hues add intriguing splashes of brightness to the forest floor.

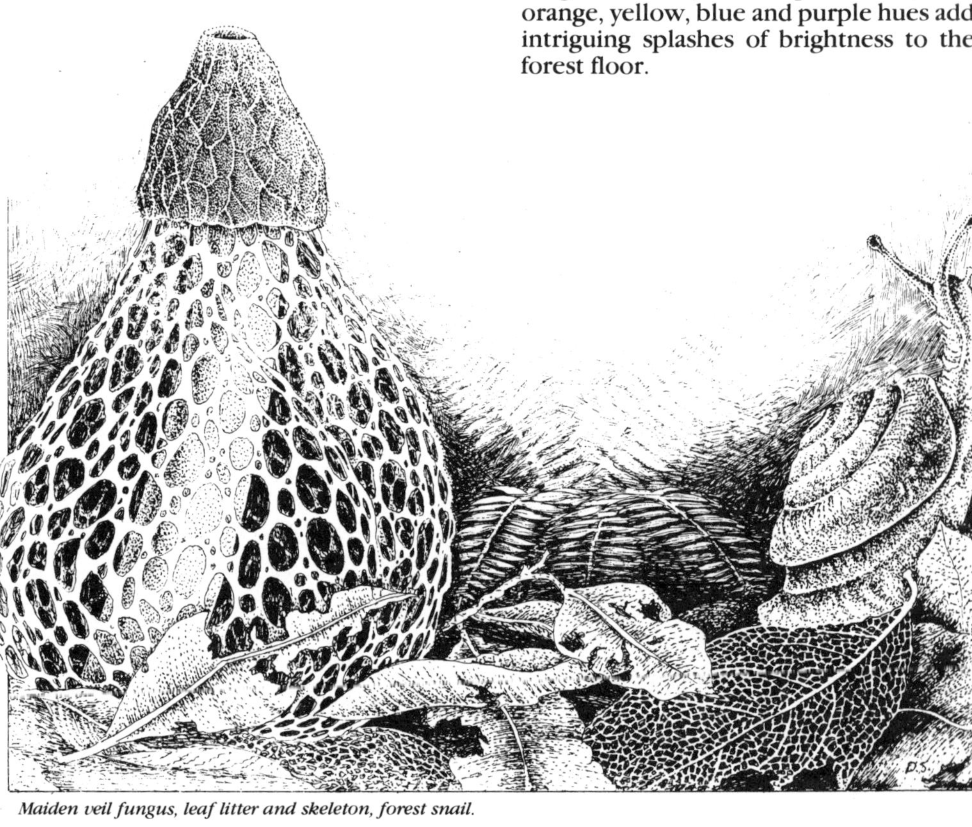

Maiden veil fungus, leaf litter and skeleton, forest snail.

The main body of a fungus, concealed within a soil or wood substrate, is called a mycelium and consists of masses of fine threads referred to as hyphae. Hyphae spread out over, or within, a rotting substrate such as leaves, logs or diseased trees and from their tips secrete digestive enzymes to break down the dead plant material. Not only do fungi obtain their own food in this way but they help to break up the rotting vegetation into still smaller bits and pieces, or detritus.

Worms, springtails, amphipods, mites and snails graze upon detritus. It is a fascinating experience to get down on ones knees and rumage through leaf litter and surface soil layers searching for these small creatures. One of the first things seen are the common litter dwelling amphipods, crustaceans closely related to sea prawns which they look much like. These small agile animals, together with flightless jumping insects called springtails, rapidly disappear from view with flea-like hops when exposed to light. Look for a leaf skeleton (page 12). Leaf skeletons, common on the forest floor, clearly indicate that the nutritious soft green parts of the leaf have been digested by the decomposers and detritivores, or detritus-eaters, leaving only the less nutritious skeletons.

Whilst walking over the forest floor it is hard to imagine this microcosm of life exists beneath ones feet — billions of micro-organisms each hectare playing one of the most important roles within the rainforest ecosystem. Whilst obtaining their own nourishment they release from the dead plant material the all-important nutrients in the form of their own body waste and, when they die, their own bodies which decompose back into the soil to be recycled by tree roots absorbing nutrients for new plant growth.

Pseudoscorpions, centipedes, beetles and other insects and their larvae feed by hunting the tiny detritivores, and in turn they themselves represent food for larger ground dwelling spiders, scorpions, frogs and lizards. This intricate and delicately balanced food web continues above ground, for litter dwelling creatures fall prey to larger carnivorous, or animal-eating, amphibia and reptiles and to birds and mammals that roam the forest by day or by night. It is these larger animals that one is most likely to see when walking through a rainforest, if doing so slowly and quietly. Drab coloured clothing will help to see the forest inhabitants too.

Grey-headed robin.

Brush turkey.

D.H.Stacey '84

The most rewarding time to visit a rainforest is at dawn, when the first rays of light filter down through the canopy and cast soft dappled light onto the forest floor. At sunrise the dawn chorus has warmed up. Whistlings of robins, such as those of the Pale-yellow Robin (see page 3) and the tropical upland Grey-headed Robin (page 13), chatterings of scrub-wrens and conspicious calls of logrunners (page 17) are some of the first to be heard. Logrunners are more often heard than seen as their plumage and cryptic behaviour effectively camouflages them against their dim forest floor backdrop. By following the direction of their calls and then, when closer, looking for moving leaf litter a flock of logrunners can be watched. Look carefully and observe their peculiar dance-like foraging behaviour. With their large feet and long claws they grasp a 'footful' of litter and hurl it aside, first to one side and then to the other, snatching up the small litter dwelling creatures with the beak. The unsuspecting exposed prey is readily spotted by the bird's keen eye-sight that is so well adapted for searching in dark situations. Whilst foraging in this manner logrunners prop themselves back onto their strong spiny-shafted tail feathers to aid their balance. These birds are indeed well adapted for a ground foraging existence.

Pairs of whipbirds, giving their characteristic whip-crack like calls, hunt the forest floor for small live prey as do the smaller Yellow-throated Scrub Wrens and, in the highlands of the Atherton Region, the rarely sighted Fern Wren. The lively Noisy Pitta also eats insects, worms and other small invertebrates as well as the occasional fruit. It is, however, particularly fond of snails which it reaches by smashing the shell onto a rock or log 'anvil'. The Noisy Pitta, unlike other ground foragers, is very brightly coloured in hues of rich buff, black, red, green and turquoise blue. Nevertheless, it is still difficult to see on the dark forest floor. A characteristic 'walk-to-work' whistle announces the pitta's presence. Try mimicking a pitta call, as it is possible to fool one into believing you are another pitta and it will then hop right up to your feet! Pittas sometimes feed their chicks on the most unloved of all rainforest inhabitants — leeches!

Rainforest dwelling leeches (below) are quite small and harmless but can nevertheless still be a nuisance. Using the sucker at either end of their body they loop their way around the forest floor and amongst leafy foliage in search of an unsuspecting host such as a bird, mammal or human being from which to suck blood. It is often advisable to carry a box of matches for burning off these unwanted hangers-on.

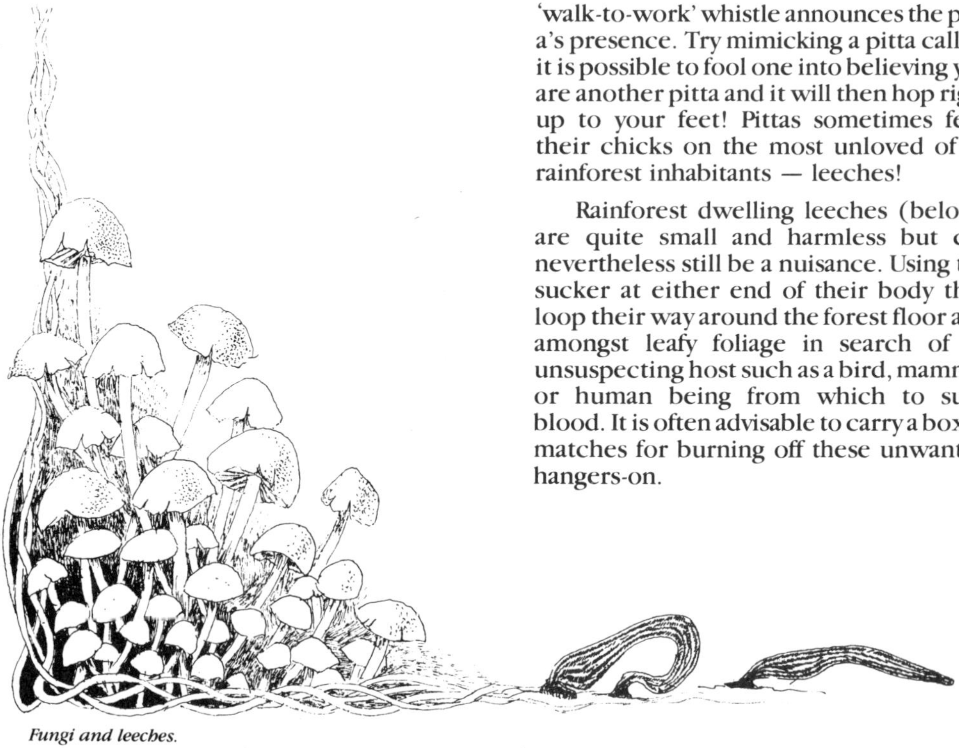

Fungi and leeches.

The Brush Turkey (pages 14-15) is perhaps one of the best known of all our ground dwelling rainforest birds for it is easily attracted to picnic tables where it readily accepts most things that are offered to it. It is in fact omnivorous, or a mixed feeder, eating animal and fruit foods alike. It belongs to a strange group of birds called megapodes, meaning large feet, or mound builders. Its large strong feet are not only ideally suited for scraping through leaf litter and rotting wood in search of foods but are also used for another purpose. During the breeding season males rake up litter and soil into a large mound within which, after mating, females lay their eggs. It is the heat generated by the compost heap vegetation that incubates the eggs, as the male regulates temperature and humidity within it by adjusting the mound mass relative to climate. When the eggs hatch the young are quite precocious and, after digging themselves out of the mound, look after themselves from the very first day of life.

The Orange-footed Jungle Fowl is another megapode, that is confined to the northern lowland tropical rainforests but unlike the wattled Brush Turkey lacks wattles and it is crested, and is predominantly herbivorous in eating fruits and leafy shoots. Its pitiful sounding loud cries, screams and wails can often be heard on dusk and may continue right through the night.

The giant flightless Cassowary bird (pages 18-19) lives within the tropical rainforests of north Queensland, as well as those of New Guinea and the Aru Islands. It eats mostly fruits, and perhaps has to wander through vast tracts of rainforest in search of enough fallen fruit to sustain its massive body weight. It is exciting indeed to meet one of these spectacular birds when walking through rainforest, as they stand as tall as an average woman and female birds, which are larger than their mates, may weigh as much as 55 kilograms. Such an experience is perhaps the Australian equivalent to coming face to face with a tiger or elephant in Asian jungles!

It is the male Cassowary that prepares a shallow nest scrape on the forest floor, incubates up to five huge green eggs for about two months and cares for the chicks (page 18-19) for approximately another nine months. Female cassowaries are the ultimate liberationsists for not only do they not take a share of nesting duties or raising of the chicks, but they may mate with and lay eggs in the nests of several males before leaving them with all the parental duties.

Male northern logrunner.

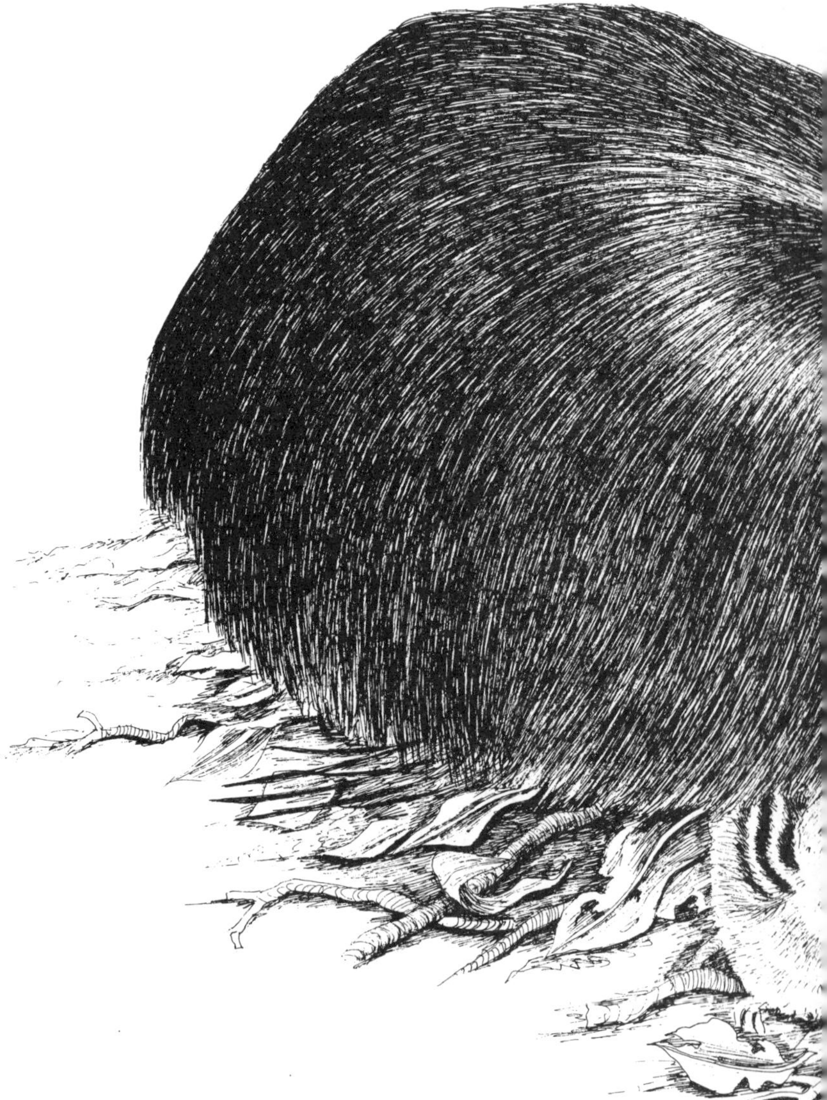

Southern cassowary — male with chicks at nest.

D.H. Stacey

During the heat of the day most animals siesta, and the rainforest becomes relatively quiet. During the months of September through to January when most birds breed, however, one bird of our northern tropical upland forests remains extremely vocal throughout the day. Listen carefully for a continuous repertoire of bewildering sounds intermingled with mimicry of other rainforest bird calls. Creep very quietly towards the source of these powerful calls, for they belong to the shy and cryptically coloured Tooth-billed Bowerbird. The male tooth-bill clears an area of forest floor of all leaves and debris to form a 'court', a simple type of bower, upon which he places leaves upside-down to expose their contrasting pale undersides (below). It is here the male calls almost continuously to attract mates, leaving only infrequently to feed upon forest fruits. Like the majority of bowerbirds, male tooth-bills are promiscuous in that instead of pairing with a female, defending a territory and raising the chicks together like most birds do, they do not become attached to any one female but mate with as many females as possible during the brief breeding season. The female alone builds a nest, incubates the eggs and raises the one or two chicks; quite the opposite to the emancipated female cassowary (page 17).

Six of the nine bowerbird species found in Australia can be seen in rainforest. The Spotted Catbird of the tropical forests and the Green Catbird of the subtropical ones are the only two Australian bowerbird species that do not build bowers, but defend nesting territories in which both the male and the female share in raising the family. In fact catbirds form pairs that may last for the lifetime of the birds. Sometimes when walking through the rainforest one may be confronted with a catbird fluttering around on the forest floor as if badly injured, its wings held open and trailing on the ground. All this distressing stumbling about is in fact the nesting catbird's distraction display, performed in order to divert you from their all-important nest or fledgling nearby!

The best known of all of Australia's bowerbirds, and the one that has led to a person who is loathe to part with anything being called 'a bit of a bowerbird' is the Satin Bower bird. The glossy blue-black male Satin Bowerbird builds an 'avenue' bower of two parallel walls of sticks which he decorates with blue and green objects, both natural and unnatural. Do not ever leave valued blue objects around where there are Satin Bowerbirds for they are likely to be stolen by a decorating male!

The Regent Bowerbird occurs in the rainforests of Eungella in the north and southwards down into New South Wales. At Lamington National Park they can be commonly seen around the picnic areas where the spectacularly plumaged adult males will feed from the hand. The striking black and golden plumaged male Regent Bowerbird builds a similarly shaped bower to that of the male Satin but it is smaller and sparser and the bower decorations are quite natural ones, consisting of yellowish leaves, snail shells and brownish seeds and berries.

Male tooth-billed bowerbird.

The world's smallest bowerbird, the Golden, occurs only in the upland tropical rainforests. It is only the size of a Common Starling but builds the largest of all bowers. When confronting one of these huge stick structures for the first time it is very hard to imagine that such a small bird is capable of such an architectural feat. The bower may consist of one or two towers and may reach as high as three metres from the ground. A display perch protrudes from the side of a single tower, or connects between twin towers, and it is at either side of this display perch that the male Golden Bowerbird places his array of decorations (below); including bearded lichen, seed pods, and various off-white flowers including wild orchids and jasmine. The sticks on either side of the display perch are more skillfully and regularly arranged than those of the rest of the bower. The bower structure is quite solid as fungi fuse the sticks together into a coherent whole.

Male Golden Bowerbirds do not attain the mature gold plumage until around six or seven years of age and until then, like the young male Satin and Regent Bowerbirds, they remain in the dullish female plumage which in the case of the Golden Bowerbird is a generally greyish appearance. It is not necessary for the females of these three species to advertize their presence in the forest and in fact it is important for them to be as cryptic as possible as they raise the young entirely on their own. Should the females be bright in their feathering they would be obvious and attract predators to their nest. The nest of the Golden Bowerbird is in fact a neat substantial cup of dead leaves and fronds built on a base of twigs and lined with tendrils. The nest is invariably built in a tree crevice or between buttresses, usually no higher than three or four metres above ground.

When a male eventually attains his bright plumage he cannot simply build a new bower for himself where he will, because he will be chased off by the permanent bower owners nearby. Instead he must wait to inherit an already occupied bower site by chasing off the established bower owner or by waiting for the owner to die. Thus, the bower sites are highly traditional, being passed on from one male generation to the next and it is conceivable that a bower site may have existed in the same forest location for many hundreds of years.

Bowers are apparently built in rainforested areas particularly rich in the fruits eaten by these birds because the bowerbirds appear to be able to spend much of their time singing, displaying, or building and decorating their bowers without having to continuously search for food. A bower owning male's foraging forays must be quite brief, particularly during the breeding season, as rival males will visit the bower during his absence and steal his decorations, or of course he may miss a female suitor. During the winter, when fruits are less abundant, bowerbirds have to spend more time searching out fruits than attending their bowers and it is during this time that they moult, or change their feathers.

Male golden bowerbird.

Dingo.

Sometimes during the day it is possible to glimpse a small mouse-like animal called a marsupial mouse, or antechinus, scurrying through leaf litter and about tree trunks hunting other small animals foods. A rather wallaby-like Red-legged Pademelon (pages 46-47) can be seen eating fallen leaves, fresh leaves, ferns, grasses and berries, although this shy animal is not usually noticed until it is hopping rapidly away with heavy foot-thumps on the ground. These and most other rainforest mammals are, however, predominantly nocturnal which is to say they are active at night.

By dusk the forest floor life changes dramatically. Logrunners, whipbirds, robins and scrub wrens enjoy their last meal of the day before flying up into the leafy foliage above to roost. Even brush turkeys crash clumsily upwards in search of an arboreal leafy haven in which to sleep safely away from ground dwelling nocturnal predators. On the forest floor the night-time blackness is here and there broken by the dull candle-light glow of luminescing fungi which, if given a very long exposure indeed with the camera on a firm tripod, can be photographed.

Sitting at a rainforested picnic area within the confines of a National Park (see page 48) is often the best way to observe some of the forest floor nightlife. Bandicoots (below) emerge from the safety of their daytime den, usually a shallow depression on the forest floor lined with leaves, and start nosing their way through litter and rotting logs for small insects and other invertebrates. Digging up grass roots in search of a tasty grub or two is a favourite pastime of these delightful creatures, much to the frustration of forest-edge garden owners. Native rats scurry around searching for fruits, seeds, insects and other small animals and will often be seen raiding litter bins for any leftover tasty morsel. The occasional Echidna (see also page 24) may be seen ambling by. The huge powerful claws of the Echidna enable it to tear apart rotting wood to reach suitable insect foods, particularly ants, which it extracts and eats with the aid of a long sticky tongue.

Dingoes (opposite) lurk quietly within the forest gloom hunting brush turkeys, bandicoots, native rats, and even catch and eat the prickly Echidna quite often.

As day breaks the night-shift retires. Roosting logrunners give their first calls in the understorey to re-establish their presence and warn neighbouring logrunners that they are again patrolling their territory. Their morning cacophony heralds in yet another day.

Long-nosed bandicoot.

CREEKS, WATERFALLS AND POOLS

To rest from walking nothing is more pleasant than to sit quietly beside a rainforest creek or waterfall. On a warm sunny day this is both a very pleasurable and productive experience, giving a sense of being as one with the rainforest and at peace with the world.

Warm rays filter down through the leafy canopy edge illuminating upper fronds of tree ferns and enhancing the intense greenness of the moss clad creekside rocks. An occasional Eastern Water Dragon may be glimpsed sunning itself on rock or overhanging branch above cascading waters. Orchids display intricate and often brightly coloured flowers (back cover) in order to attract suitable insect pollinators such as wasps, bees, moths and butterflies. Some flowers even mimic the shape of their specific insect pollinator in order to attract them to their nectar and so ensure pollination. The filmy ferns, fan ferns, maiden veil ferns and violets growing along wet banks and between rock crevices clearly emphasize the incredible wealth of plant life in this damp environment.

Rainforests cannot exist without high annual rainfall. In the wet tropics this may reach 3 to 4,000 millimetres or even more a year, most of which falls in the summer months of December through to March — the wet season. During these months the relative humidity within the forest may often reach saturation point. Swirling cloud mists add to the extreme dampness. It is said that in virgin rainforest, now rare in Australia, less than half the rain that falls penetrates through the canopy to the forest floor. Some rainforest trees have evolved drip-tips at the end of their leaves to channel off excess water. Below the canopy the rain is absorbed directly by tree bark, some epiphytic structures, and some evaporates, whilst the remainder does eventually indirectly reach the forest floor. Excess water drains off and down into mountain gullies and into creeks to eventually cascade over waterfalls and flow into lowland rivers and onwards to the sea.

A whole host of animals rely on this all-important water supply not just to drink or bathe in, or to reproduce in as frogs do, but for food supplies. Within the water itself exist myriads of micro-organisms and larger aquatic animals. Innumerable microscopic invertebrates, water insects and their larvae, yabbies or freshwater crayfish, tadpoles, eels, colourful fish, turtles and platypus are just some of the permanent residents, whilst frogs and water snakes are regular visitors.

The Platypus and the Echidna are the only two representatives of the monotreme or egg-laying mammals, the most primitive of the three major groups of mammals (marsupials and placentals being the other two). Platypus burrow into creek and lake banks, usually under a protective tangle of roots. Although most active at dusk or dawn it is sometimes possible to observe one paddling around a deeper forest creek pool during daylight. Their webbed fore feet propel them through the water whilst the hind feet,

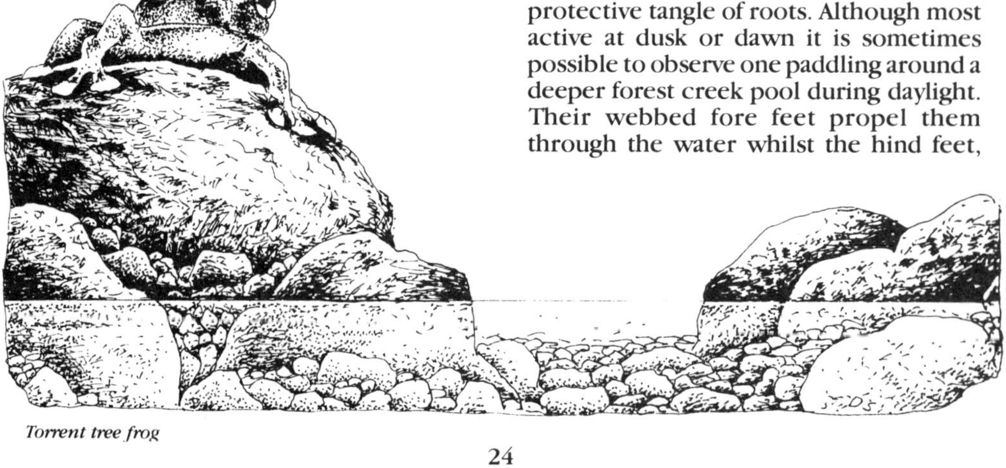

Torrent tree frog

although webbed, are held against the tail to enhance streamlining except during steering and braking. The large soft beak of the platypus is well adapted for snapping up large food prey or sifting through mud and extracting invertebrate foods that the touch-sensitive bill detects. When diving a platypus closes its eyes, nostril and ear aperatures, and captured food is stored in large cheek pouches to be eaten above water later.

Sunlight glinting on the vibrant blue plumage of an Azure Kingfisher may reveal its presence on a nearby rock or branch. A kingfisher will perch motionless, save for sharp vertical head bobs, for prolonged periods whilst peering down intensely into the water below in search of a tasty frog or fish. Suddenly and efficiently it dives down into the water to capture the unwary prey, returning to a nearby perch to beat it to death and devour its catch head first.

Frogs are the commonest animals associated with water in rainforest, for it is here during their breeding season that most must congregate and mate. Some, such as the Northern Great Barred Frog of tropical forests, visit water only to breed spending the rest of their time in damp situations throughout the forest floor; whereas the tropical rainforest dwelling Torrent Tree Frog remains close to fast flowing creek waters. Like all tree frogs, the Torrent Tree Frog has almost circular toe pads on the end of each of its digits that act as efficient sucker pads holding the frog securely to slippery surfaces. Disturb this frog and it will leap into the swirling white waters to escape you.

It is the Red-eyed Tree Frog (page 49) that is perhaps the best known of all our rainforest frog species. On a warm and wet summers night these brilliant green Red-eyed Tree Frogs, unseen and unheard for months during the winter, congregate amongst low leafy vegetation close to small pools and creeks to commence their loud vocalizations. It is often difficult to see these appealing green creatures because they blend in so well with their background, but their calls and certainly their striking bright red eyes ultimately give them away. By inflating and deflating their bright yellow vocal sacs they emit loud moans followed by quieter trilling sounds. Other species of frog often join in, along with the repetitious, often irritating, calls of crickets to form a nocturnal choir.

Some people think themselves into a phobia about frogs, but these splendid and appealing animals are utterly harmless. In fact the Green Tree Frog (page 27) that commonly lives around habitations is very beneficial to the homestead as flies and spiders are an important part of its diet. A remarkable frog discovered as recently as January 1984 in the small mountain top rainforest creeks of Eungella National Park in Queensland may prove to be of immense value to medical science. Females of this Gastric Brooding Frog swallow their eggs upon laying them and retain them in their stomach, not eating in the interim, until the eggs become tadpoles and subsequently develop into froglets. The female then returns to water to spit her young into the stream where they go their own way. To carry out this remarkable feat the frog must obviously produce a chemical in her eggs or gut that inhibits digestion. This chemical may be of immense value in treating human gastric ailments, like ulcers.

During the heat of the day frogs seek out moist refuges in which to hide to prevent them becoming too dry. It is most important that these animals keep moist for they breathe through their skin as well as their lungs. During extreme cold frogs will hibernate in the deeper warmer layers of rotting vegetation. A good spot to find them is beneath rotting litter that has accumulated on the base of a large Crow's Nest Fern (see page 28), not necessarily close to water but quite often high up in the upper forest zones, as long as it is constantly damp.

Green tree frog.

THE UNDERSTOREY

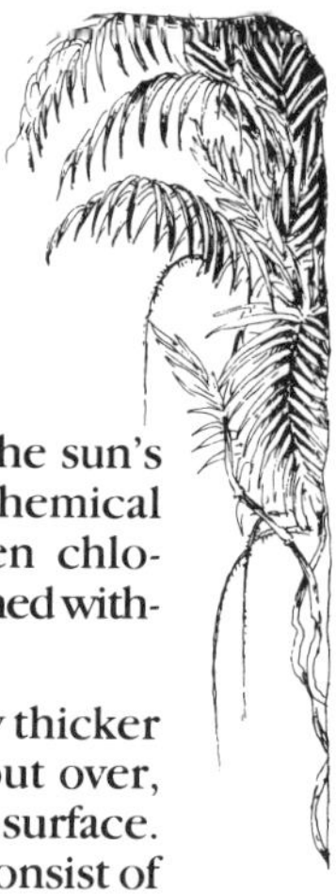

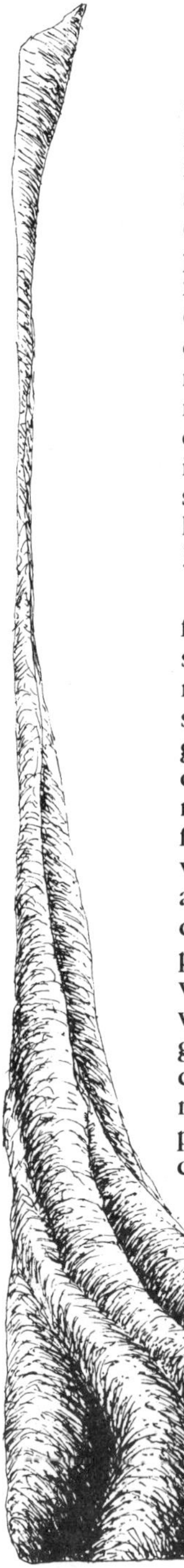

It is the forest understorey (opposite) one sees most of when walking through a rainforest. It is this zone, between forest floor and canopy, that supports an immense wealth of plant life. It is hard to comprehend that within our rainforests there are as many as 600 to 800 different species of large rainforest tree, and new species are still being regularly discovered and described. Besides trees there are innumerable varieties of ferns, shrubs, palms, climbers, stranglers and epiphytes (see page 30).

In the dark gloom of the forest floor delicate seedlings struggle for survival (below right), competing for light, space and nutrients with the ground orchids, conjevois, cordylines and wild gingers to name only a few. The successful saplings grow slowly upwards into the understorey and on ever-upwards into the canopy to reach that all-important energy giving light which is imperative for survival to maturity. Without light, green plants are unable to carry out photosynthesis, the means by which some of the plants' food is built up from carbon dioxide and water using the energy from the sun's rays, by a series of chemical reactions in the green chlorophyll pigment contained within the plants' leaves.

Sapling roots grow thicker and begin to spread out over, or just below, the soil surface. Most rainforest soils consist of a very shallow layer overlying thick red clay and so plant roots must grow near to, or on, its surface where most nutrients are concentrated.Some roots extend out directly from the trunk base, but other form buttresses at the base of the tree. Buttresses (below left) are a striking feature of some rainforest trees and comprise woody plank-like extensions radiating outwards from the lower part of the tree base. These structures, which may reach enormous proportions around the base of a rainforest giant, provide some support and stability to the tree. Standing between the huge buttresses of a rainforest giant makes one feel very insignificant indeed, especially if one might be looking up at a tree that has stood on this earth for 2,000 years or more.

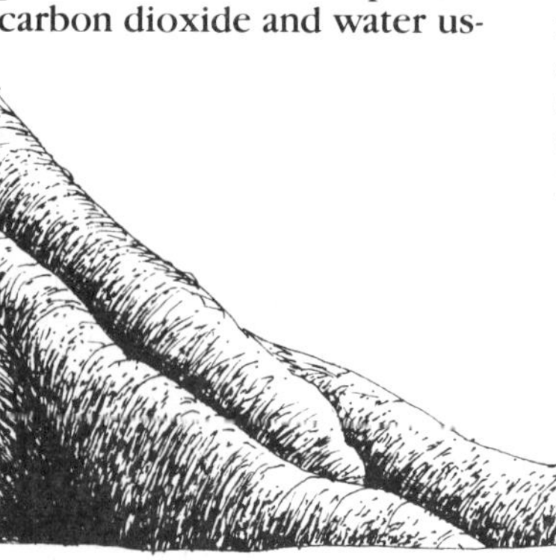

Buttress, seedlings.

Lawyer vine.

Strangler fig, crow's nest and staghorn ferns.

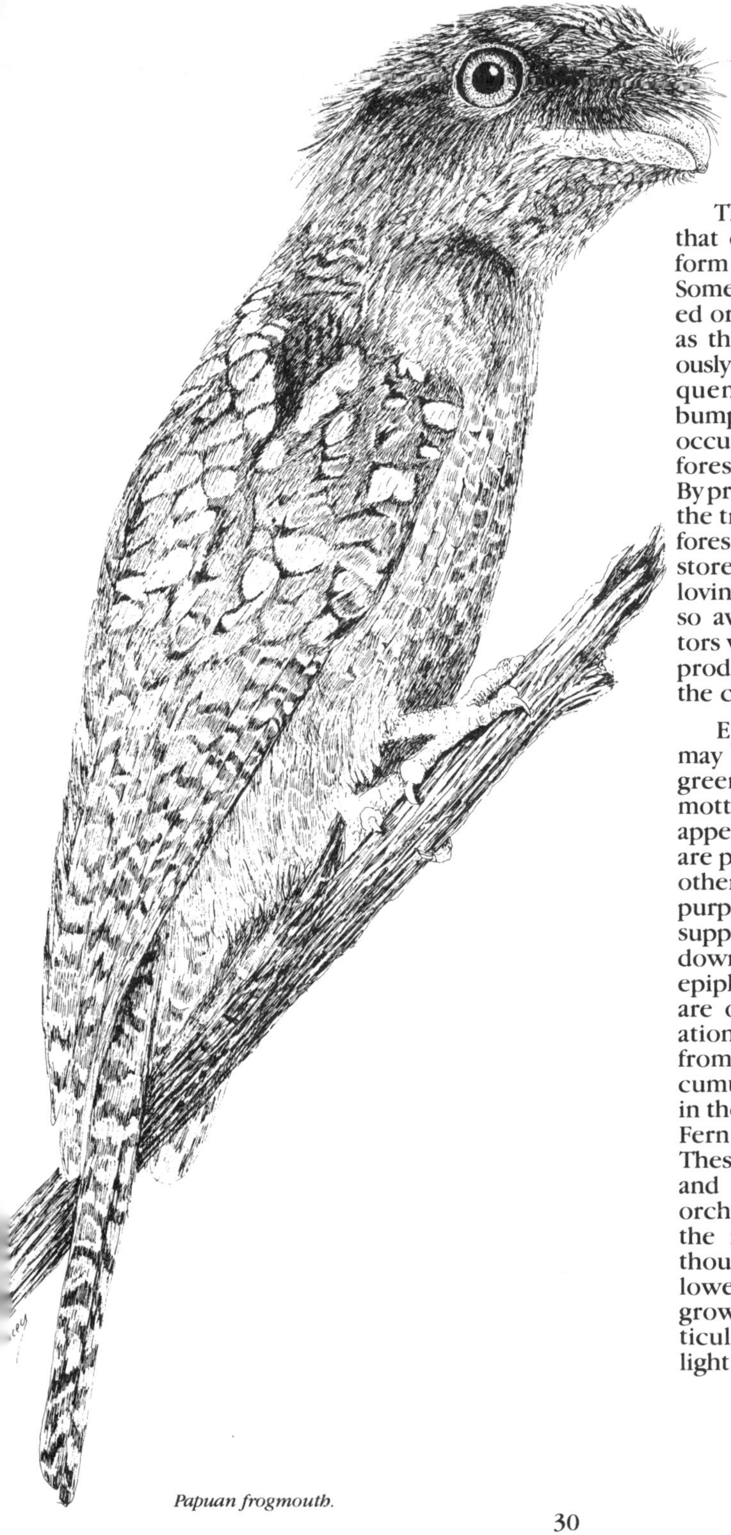

The hundreds of tree trunks that one sees in the understorey form all shapes, sizes and colours. Some are quite smooth, other ridged or furrowed whilst some, such as the Bumpy Satinash, are obviously bumpy. Flowers and the subsequent fruits grow from these bumps. This strange phenomenon occurs in several species of rainforest tree and is called cauliflory. By producing flowers directly from the trunk, often right down to the forest floor, the tree attracts understorey pollinators such as shade-loving butterflies and moths and so avoids competing for pollinators with the majority of trees that produce their flowers high up in the canopy.

Epiphytic mosses and lichens may decorate bark with hues of greens and yellows giving it a mottled and delicately textured appearance (opposite). Epiphytes are plants that use tree trunks and other vegetation for attachment purposes; they do not harm their support in any way. Rain trickling down tree trunks is absorbed by epiphytic roots and their nutrients are obtained from rotting vegetation trapped in bark crevices, or from fallen leaves that have accumulated on top of their roots as in the larger Crow's or Bird's Nest Fern, Elkhorn and Staghorn Ferns. These ferns (page 29 top right) and some of the tree-dwelling orchids (back cover) are some of the most familiar epiphytes. Although they normally reside in the lower part of the canopy some grow within the understorey particularly where it is a little more light.

Papuan frogmouth.

Epiphytic ferns may reach massive proportions, and provide an excellent catchment area for falling leaves and other debris. This rotting plant material not only provides these ferns with nutrients but also supports a miniature community of typically forest floor dwelling litter life. Insects and their larvae, amphipods and other small creatures thrive in these high situations along with frogs (see page 26) and reptiles that shelter within the foliage of the epiphytes. It is not unusual to see the normally ground foraging whipbirds searching for food at these great heights, because here they find a bit of forest floor up in the air! Some birds such as catbirds (page 20 and 41), rifle birds (pages 42-43) and Brown Pigeons may select the platform formed by the top of these basket-shaped epiphytic ferns to build their nests.

Not all ferns are epiphytic by any means. There is a wide variety of smaller ferns growing on the forest floor and particularly along creek banks as well as the conspicuous and elegant tall tree ferns. Tree ferns favour sunny situations such as along creeks or where there is an unnatural break in the canopy such as a track. Their fronds spread outwards umbrella-like to receive maximum light. The underside of each frond at times bears thousands of small spores held in symmetrically shaped brown structures by which different fern species can be identified. A fern spore germinates in damp conditions to form a tiny heart-shaped leafy structure called a prothallus. It is this that then produces the egg and sperm cells that eventually fuse to give rise to the fern plant as we know it.

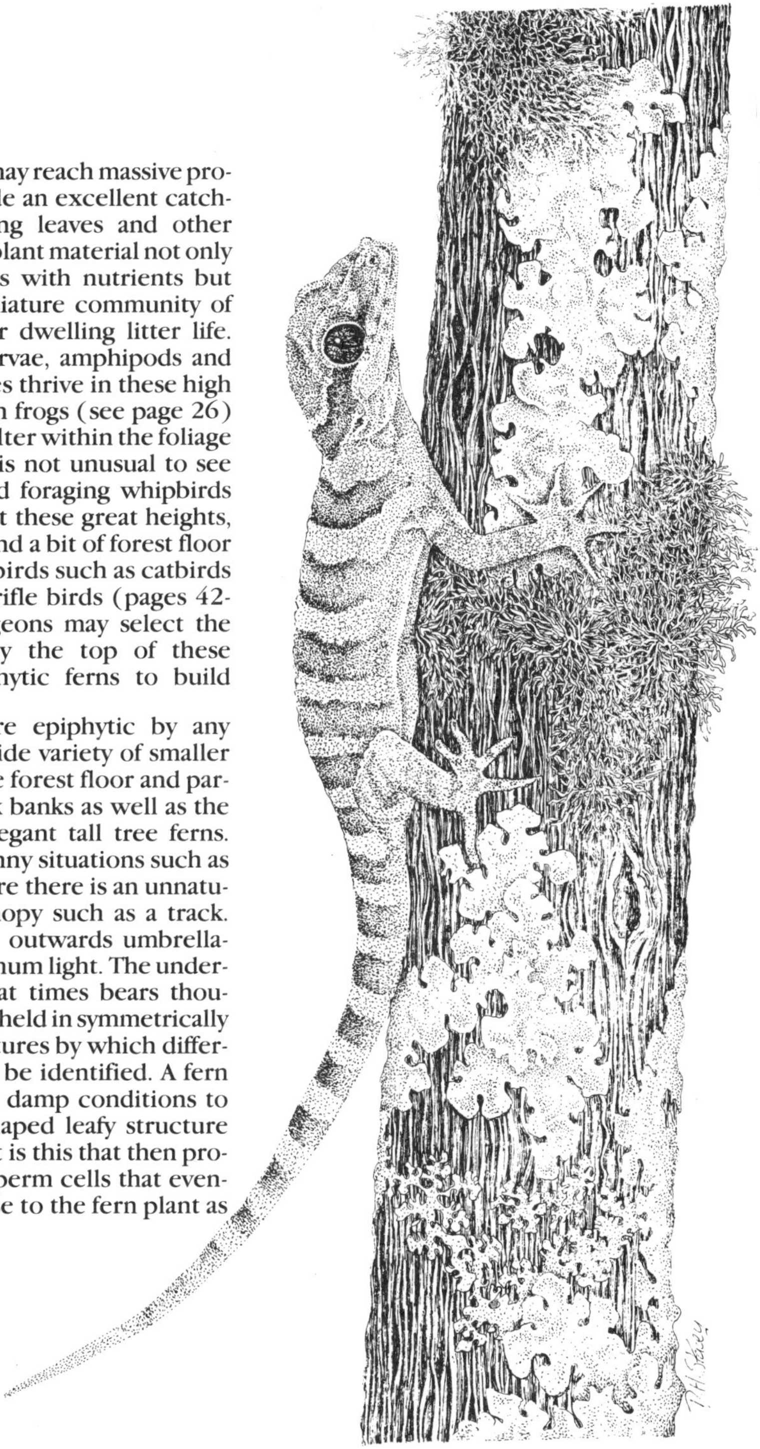

Ring-tailed gecko, on moss & lichen-covered bark.

The Bangalow Palm, Fan Palm, and Walking Stick Palm are just a few of the palm tree species associated with Queensland rainforests. Not all palms, however, are trees that stand on a trunk. A walk in the rainforest would be incomplete without getting tangledup in the most familiar of all rainforestclimbers, the Lawyer Vine or Wait-a-While. Lawyer vine is in fact a true but trunkless palm that has evolved a means of climbing up to that all-important light source in the canopy. It puts out long tendrils armed with recurved hooks (page 28 top right) from the apex of the plant and these latch onto adjacent foliage and support the climbing vine as it makes its way ever-upwards. Leaves and stems of Lawyer Vine are armed with needle sharp spines which not only protect the palm itself from being eaten but provide well protected nesting sites for many of the understorey dwelling birds such as robins, scrubwrens and shrike thrushes. It is difficult indeed for a predator such as a native tree-climbing rat (above) to steal eggs or chicks from a bird's nest.

Monsteras, phyllodendrons and pandanus vines anchor themselves firmly to a tree trunk by means of small modified lateral roots that extend from their main climbing stem and attach to a tree. Supple Jack or Flagellaria vines have spring-like tendrils on their leaf tips for coiling around a suitable support. Most climbers, however, do not have such sophisticated aids to climbing but simply spiral or entwine their way up a trunk into the canopy. Some of these become thick and woody to form lianes. Having reached the canopy lianes then travel horizontally making their way through the leafy forest canopy from tree crown to tree crown to great lengths. Stems may reach massive proportions, some round and twisted and others wide and flattened.

The mighty Strangler Fig (see page 29) actually kills its host tree. The fruits of the well known strangler fig tree are eaten by canopy dwelling birds and mammals and the indigestible seeds are voided and

Fawn-footed melomys.

dispersed. These seeds germinate high up in the canopy in some crevice or tree hole. The seedling puts out long thin 'cable-like' roots which make their way down the host tree to the ground where an ample supply of water and nutrients is available. These roots then more rapidly expand to entwine and fuse, forming a strangle-like hold around the trunk of the tree. Eventually the host trees growth is restricted and it is strangled by the fig, the canopy of which has replaced that of the host tree. The host tree slowly rots away leaving a free fig tree that may live for many hundred of years.

Robins, fantails, scrub-wrens, warblers, shrike thrushes and tree creepers live mainly within the understorey where they feed upon the wealth of insects and other invertebrates that abound there. Cicadas, moths, butterflies, bees, wasps, ants, flies, preying mantids, stick insects (page 50) and spiders are just some of the understorey residents.

It is hard to believe that with such a veritable abundance of food there would be any competition for it by these small birds. Nevertheless insect-eating birds have gone ut of their way to avoid competing with each other by evolving different bill shapes and sizes, by foraging at different heights or sites, or by performing different behaviours. It is fascinating to sit and watch these small birds feeding. The Grey Fantail tends to forage higher than its close relative the Rufous Fantail and so avoids competing with it to some extent. Trunks and branches are favoured foraging sites of tree creepers whereas shrike thrushes methodically hunt foods from leafy foliage. Robins invariably peer down to the ground as if mesmerized, searching for moving prey in the litter below before diving down to capture it, whereas the Large-billed Scrub Wren employs a searching behaviour to locate insects and spiders from within tangles and dead leaves.

Watch tree trunks carefully for any sign of life. Besides tree creepers, the occasional tree monitor, skink or rainforest dragon (below) might be spotted stealthily hunting insects up and down the trunks. During the night these diurnal reptiles hide away in tree hollows, holes or beneath bark, and whilst they rest nocturnal counterparts emerge from their similar retreats to start their night-time forays, perhaps even to prey upon those resting.

Boyd's rainforest dragon

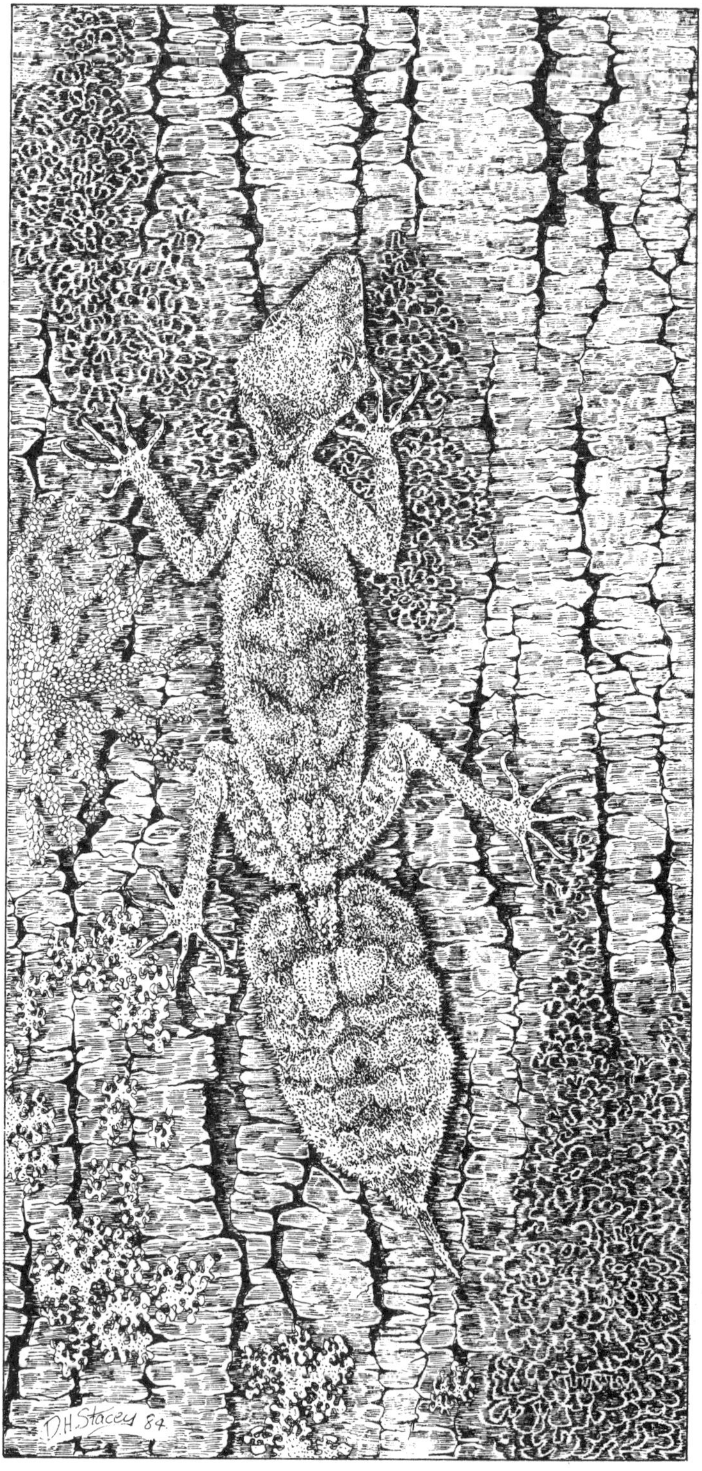

Northern leaf-tailed gecko.

During the day the remarkable Northern Leaf-tailed Gecko (opposite) hides away beneath loose bark against which it blends in perfectly. Pressing its green and brown pigmented and irregularly outlined body against a tree trunk it eliminates its shadow and makes itself almost invisible. It is a classic example of cryptic or adaptive colouration. Even the irregular pigmentation on its iris enhances this near-to-perfect camouflage. As its name suggests it has a large leaf-shaped tail which is quite fleshy and which the gecko moves about when disturbed. Because its tail looks somewhat like a head the movement fools the predator into thinking it is grabbing its prey's head! In fact when caught the gecko sheds its tail, which then continues to thrash about and the gecko escapes to safety leaving the predator busy with the lively tail. Later the gecko grows a new tail. Unlike most other geckos, that have pads on the end of their digits to assist movement over slippery surfaces, the Leaf-tail has claws for better traction over the ridged and furrowed bark surfaces on which it hunts for insects.

There is a tendancy to think that rainforests house particularly dangerous snakes in great abundance whereas, compared to those dwelling in drier habitats such as the Death Adder, Taipan and Brown Snake, rainforest serpents are very predominantly harmless. Tree snakes and pythons are common in the understorey. The Brown Tree Snake is in fact midly venomous but is not considered harmful to man and the other tree snake species are quite harmless. Tree snakes prey upon small lizards, birds and mammals. During the day, if you look for them, it is not unusual to see a python such as the Scrub Python of tropical forest (right), coiled up in a sunlit area sunning itself as, like all reptiles, pythons are cold blooded and must warm up before they can become active. Pythons detect their warm-blooded prey such as birds and mammals by heat-sensitive tissue contained within pits along their lower jaw. Once captured in their jaws they constrict, or strangle, their prey to death. By dislocating its jaws a python can easily swallow an animal such as a bandicoot (page 23) or large rat of a diameter much greater than that of its own body.

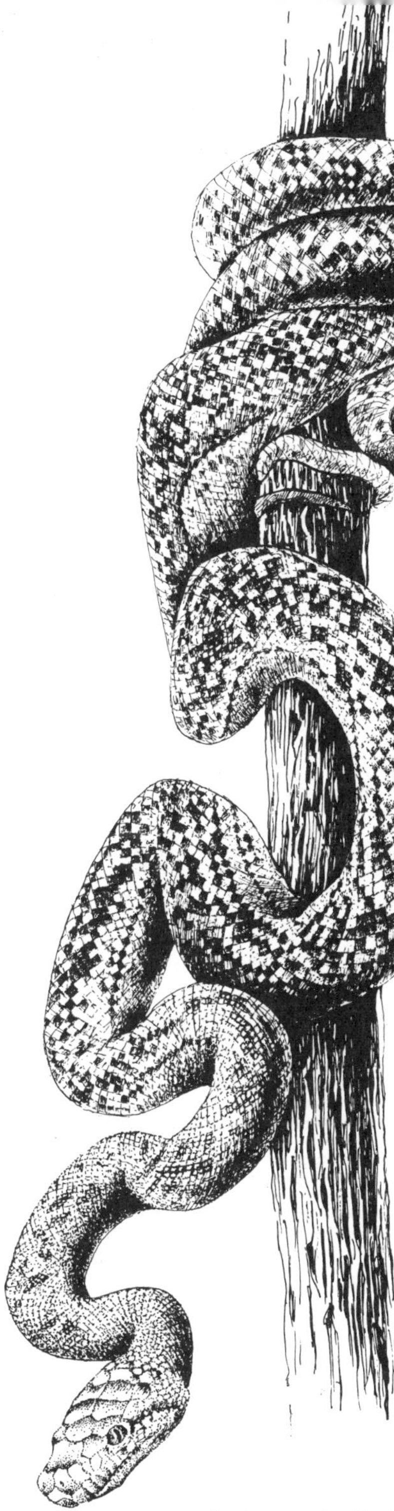

Scrub or amethystine python.

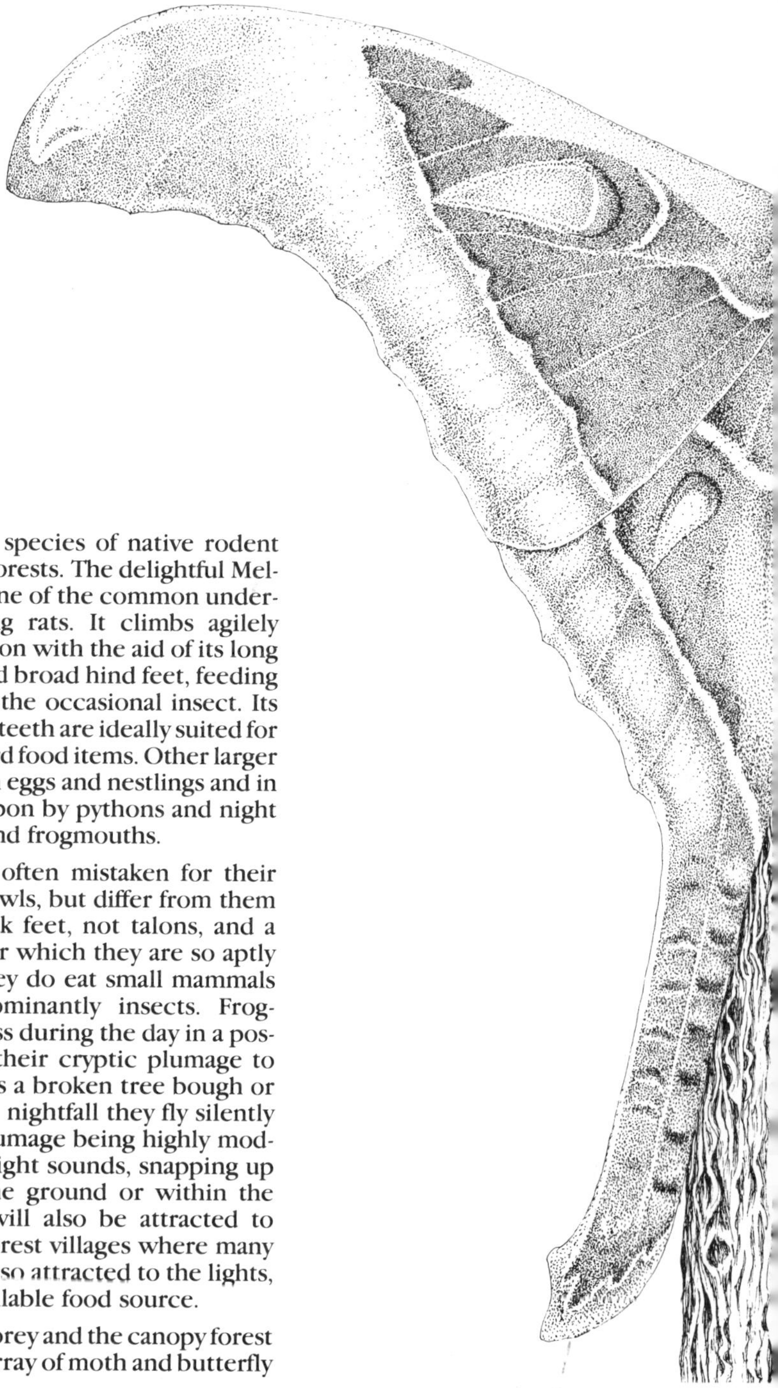

There are many species of native rodent rats within our rainforests. The delightful Melomys (page 32) is one of the common understorey tree climbing rats. It climbs agilely through the vegetation with the aid of its long shiny hairless tail and broad hind feet, feeding on fruits, seeds and the occasional insect. Its sharp rodent incisor teeth are ideally suited for gnawing through hard food items. Other larger native rats feed upon eggs and nestlings and in turn are predated upon by pythons and night birds such as owls and frogmouths.

Frogmouths are often mistaken for their close relatives, the owls, but differ from them in having small weak feet, not talons, and a large wide mouth for which they are so aptly named. Although they do eat small mammals their diet is predominantly insects. Frogmouths sit motionless during the day in a posture that enhances their cryptic plumage to make them appear as a broken tree bough or stump (page 30). At nightfall they fly silently around, their soft plumage being highly modified to muffle any flight sounds, snapping up animal prey from the ground or within the understorey. They will also be attracted to street lights in rainforest villages where many beetles and moths, also attracted to the lights, provide a readily available food source.

Both the understorey and the canopy forest zone support a vast array of moth and butterfly

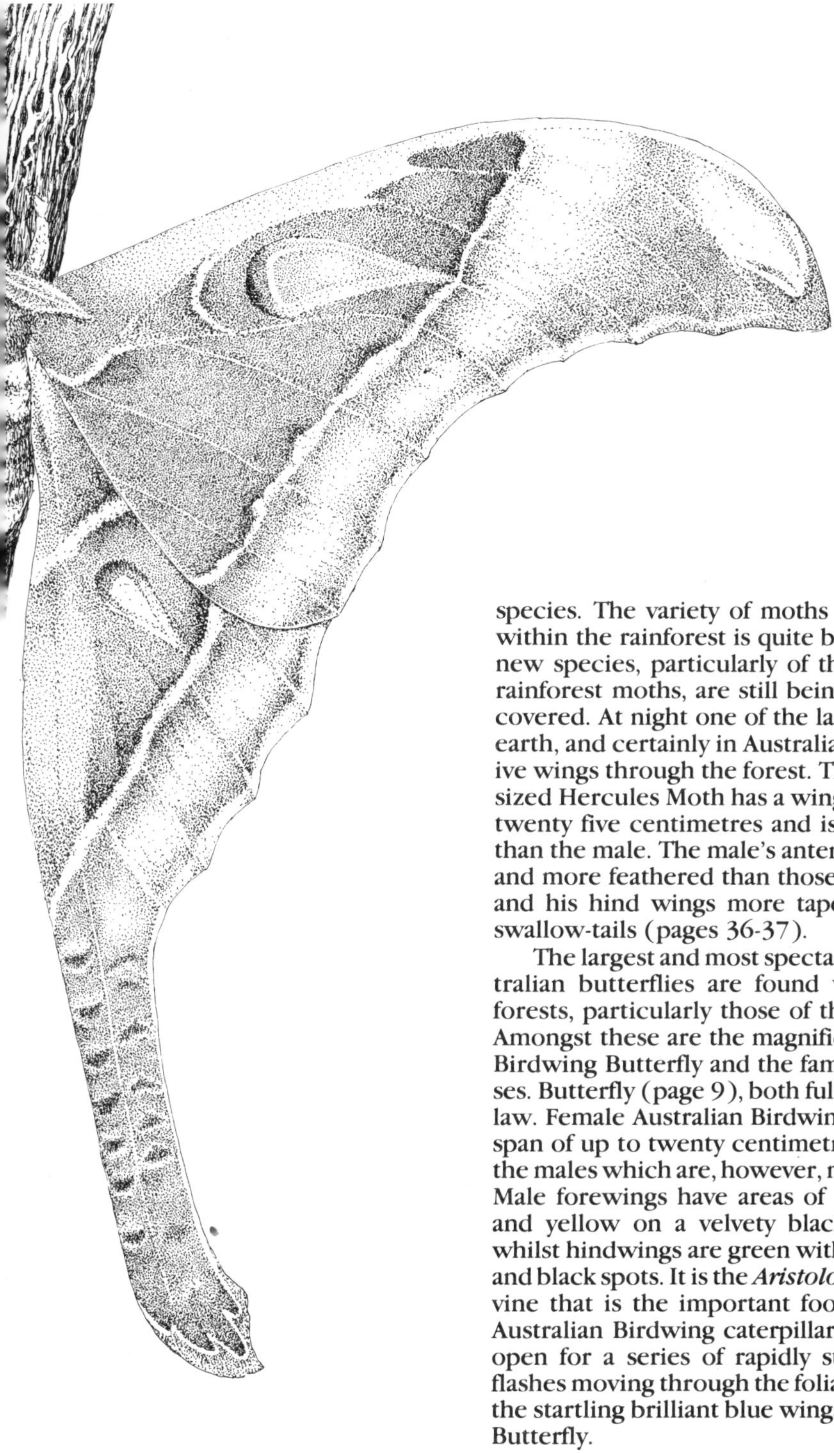

species. The variety of moths and butterflies within the rainforest is quite bewildering and new species, particularly of the little known rainforest moths, are still being regularly discovered. At night one of the largest moths on earth, and certainly in Australia, flaps its massive wings through the forest. The female bird-sized Hercules Moth has a wing span of about twenty five centimetres and is slightly larger than the male. The male's antennae are longer and more feathered than those of the females and his hind wings more tapered into finer swallow-tails (pages 36-37).

The largest and most spectacular of all Australian butterflies are found with our rainforests, particularly those of the wet tropics. Amongst these are the magnificant Australian Birdwing Butterfly and the famous Blue Ulysses. Butterfly (page 9), both fully protected by law. Female Australian Birdwings have a wing span of up to twenty centimetres, larger than the males which are, however, more colourful. Male forewings have areas of metallic green and yellow on a velvety black background, whilst hindwings are green with bright yellow and black spots. It is the *Aristolochia* rainforest vine that is the important food plant of the Australian Birdwing caterpillars. Keep an eye open for a series of rapidly successive blue flashes moving through the foliage as these are the startling brilliant blue wings of the Ulysses Butterfly.

Male hercules moth.

THE CANOPY

Interlocking tree tops, clad with beautiful flowers and colourful succulent fruits, supporting an arboreal garden of tangled vines, mosses, lichens, orchids, mistletoes (see page 40), and giant epiphytic ferns, constitute the rainforest canopy. Some trees, the giant emergents, reach impressive heights projecting well above, and spreading their massive crowns over, the canopy. From below it is almost impossible to comprehend the magnitude of the canopy trees or the intricate complexity of this arboreal community. How fantastic a canopy board-walk would be through the tree tops with a platform atop some rainforest giant to allow one to be intimately part of, and observe, life within that green sea of leaves. To be able to enjoy the brilliant profusion of the canopy flowers providing outstanding splashes of colour against a rich green background. One of the most spectacular of all rainforest flowers is the vivid red Wheel-of-Fire. This tree belongs to the family Proteaceae which includes many of the other well known forest trees such as the macadamias, grevilleas and other silky oak trees.

Imagine seeing a resplendent male Rifle Bird (pages 42-43), a true bird of paradise, displaying his glossy velvety blue-black metallic plumage in full sunlight high atop the exposed tree stump of an emergent tree. When a female arrives on the scene the male throws his peculiarly rounded wings up either side of his upstretched neck and head, and sways his body from side to side whilst bobbing up and down and rapidly flicking his head from one raised wing edge to the other. Occasionally he opens his bill to expose the brilliant yellow interior of his mouth. All to impress the ladies!

The undisturbed canopy provides a rich source of food for its residents. Butterflies, bees, lorikeets and honeyeaters feed on the nectar-yielding flowers during the day and in so doing act as important pollinators; and at night moths, blossum bats and, in the northern upland tropical rainforest the Long-tailed Pygmy-possum, fill this important role. The bill lengths and shapes of honeyeaters have evolved to enable the birds to tap the nectar source of all different flower structures. For example the Eastern Spinebill (page 51) has a long thin tapering bill which is ideally suited for probing deep down into the tubular-shaped flowers. That of Lewin's Honeyeater is much stouter and this bird, besides nectar, also includes fruits and insects in its diet. At certain times of the year when flowers are in abundance non rainforest-resident honeyeaters such as the Scarlet and the White-cheeked Honeyeaters often move into the rainforest from drier habitats to enjoy the rich nectar supply, thus further ensuring the pollination of rainforest trees.

The succulent rainforest fruits are eaten by bowerbirds (pages 20-21), rifle birds (pages 38, 42-43), fruit pigeons, possums (pages 43-44) and flying foxes. Several pigeon species occur within our rainforests, such as the Purple-crowned (below), Red-crowned, Wompoo, White-headed and Topknot Pigeons. Unlike those of drier habitats, rainforest fruit pigeons or doves are quite colourful and some even spectacularly so. It is easy to hear these birds calling but far from easy to spot them despite their bright colours. 'Ooms' and 'coos' and rumbling grunts echo through the canopy and the fabulous Wompoo Pigeon adds its well known 'wollack-a-woo' or 'wam-poo' calls to the tree top chorus.

Purple-crowned pigeon.

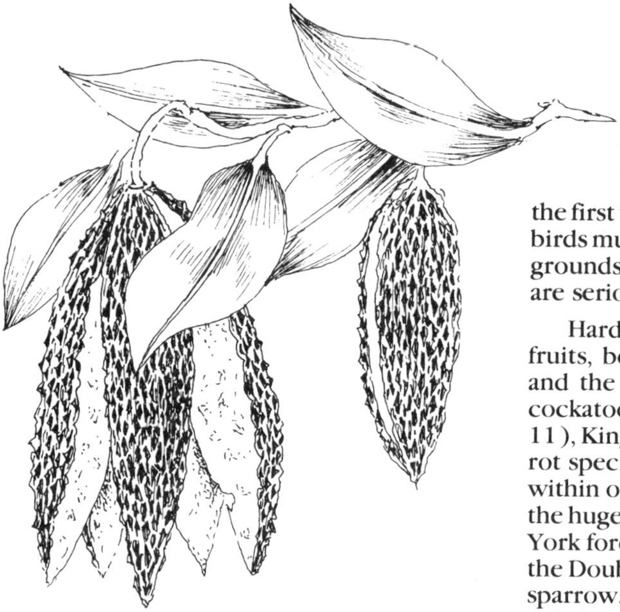

Pigeons build very flimsy nests of a few short sticks which they place upon a branch fork, on a frond of a tree fern, or among vine tangles. It is often easy to see a pigeon's single egg or young right through the sparse nest from directly below. Beware of disturbing these birds, however, for they are extremely shy and sensitive at the nest and easily desert it altogether. During the non-breeding season when foods may be scarce flocks of pigeons are often glimpsed flying from one rainforest area to another, sometimes from uplands to lowlands or vica versa, in search of richer fruit crops.

Not all of the fruit pigeons live permanently in rainforest. Everyday at dawn numerous flocks of dozens to hundreds of the tropical Torres Strait Pigeon leave their island or coastal mangrove roosting and breeding sites to forage within lowland rainforests. It animals like these that have is been so badly effected by the continual logging of giant fruit-bearing trees from the few remaining stands of virgin lowland rainforest. Needless to say logged forests have a much depleted canopy as it is the older rainforest giants, whose mature crowns support far greater fruit crops and associated plants than do smaller trees, that are the first to be felled. What a dilemma these birds must face if their traditional foraging grounds are suddenly no longer there or are seriously depleted.

Maple silkwood tree seed pods.

Hard woody nuts and seeds, succulent fruits, berries, buds, leafy shoots, nectar and the occasional insects are eaten by cockatoos, Crimson Rosella parrots (page 11), King Parrots and other colourful parrot species. A wide variety of parrots live within our rainforest, varying in size from the huge Palm Cockatoo of northern Cape York forests to Australia's smallest parrot, the Double-eyed Fig Parrot, smaller than a sparrow. Continuously falling bits of tree debris will occasionally stimulate one to look upward into the canopy and discover the cause, usually a flock of feeding parrots. Once a flock has located a tree laden with suitable nuts or fruits it has a tendency to methodically, and sometimes wastefully, work its way through the tree until the crop is exhausted which may take several days or more. Their strong massive bills are ideally suited for crushing up particularly hard food items and is specially hinged to the skull to allow maximum leverage, and thus efficient crushing of the item in 'hand'.

These fruit and seed eating canopy dwelling animals are important dispersal agents for the rainforest trees. So important is it for a tree to have its seeds dispersed that some tree species have evolved ways of fooling the forager. Fruits of such trees are often large, colourful and succulent-looking so that an animal believes they are a tasty meal. In fact the only food it gets from such a fruit is a thin layer of outer pericarp, within which is a large indigestible seed. Thus, the tree ensures its ultimate survival as the bird or mammal will void the seed elsewhere in the forest. Interestingly it has been shown that some seeds are in fact more viable once they have passed through the digestive system of a fruit-eating bird or mammal.

Mistletoes have evolved a fascinating way of ensuring their dispersal. Mistletoebirds and other frugivores, or fruit-eaters, eat the soft mistletoe berries. The seeds within are sticky and when voided adhere to a canopy branch and germinate there. Mistletoe is a partial parasite, that is it absorbs water and nutrients directly from the host tree to the tree's detriment, but the mistletoe's green leaves carry out photosynthesis (see page 28), as do all green plants, to produce additional foods. Rainforest mistletoe flowers are particularly colourful, unlike the more familiar ones we associate with Christmas but the plants themselves can be seen in sparser tree foliage as a dense clump of finer leaves.

Spotted catbird.

An experience in rainforest not to be missed is spotlighting at night for possums. For those unprepared to walk through the forest at night, spotlighting around forested picnic areas or camping sites is often just as rewarding. During the day possums hide away in tree hollows, deep within the baskets of epiphytic ferns, or within leafy dens. They emerge on dusk, or later, to commence their night-time foraging activities. Ringtails and gliders, brushtail and cuscus, and striped and pygmy are just some of the varieties of possums that can be found within different areas of rainforest. It is the widely distributed Common Ringtail or Common Brushtail that most people are familiar with, as these delightful creatures often live in close association with human dwellings. In upland tropical forests the normally greyish Common Brushtail sports

a more coppery coat and is called the Coppery Brushtail, but it is the same species (page 44).

The greatest diversity of possum species occurs within the northern tropical rainforests and it is in these areas that keen possum watchers congregate to observe and learn about their habits. Lemuroid, Herbert River, Daintree River and Green Ringtail Possums occur only in upland forests of the general tableland area (see page 48) and nowhere else in Australia. Their diet consists predominantly of leaves. Quite unlike them in diet is another northern endemic, the appealingly black and white striped Striped Possum, which eats mostly insects such as large beetle larvae which it hunts out from tree trunks, limbs and even rotting logs on the ground. It uses its two very long lower incisor teeth to gouge open bark to expose prey, which it then extracts either with its long tongue or, if well embedded, with the sharp claws on the end of each of its elongated forth fingers.

Only in the northern forests of the Cape York Peninsula can the two strange cuscus species be seen, as they occur nowhere else in Australia although also found in New Guinea and islands. It is these animals, as well as other species such as the rarely seen Green Python, Yellow-billed Kingfisher, Eclectus Parrot and Palm Cockatoo, that only occur on Cape York as well as New Guinea, that clearly illustrate the close relationship between the Australian and New Guinean rainforest faunas. Past sea level changes exposed land bridges between the land masses and so there was a flow of animals from New Guinea down to Cape York, some of which remained isolated within the forests there. It should be stressed, however, that Australia's rainforests are unique with respect to a large number of its plants and animals; very many of which are endemic, or peculiar, to our forests between Cooktown and Eungella. A clear indication of how little we know even in these very accessible and well visited southern tropical rainforests is the fact that at Eungella a new distinct honeyeater was discovered as late as 1977, not to mention the Gastric Brooding Frog (page 26).

Victoria's rifle bird — male (left) and female.

Coppery brushtail possum.

When spotlighting the first thing one usually sees of the animal is two large red eyes, that are so well adapted for night time vision, staring back down at one from leafy tree tops. Often a possum sits motionless, as if hypnotized by the light, looking directly back at the observer. It is a delight indeed to see a young possum riding piggy back on its mothers neck or back. The youngster of a Green Possum clings so tightly onto mum's back that occasional handfuls of fur are pulled out and her coat becomes progressively tatty as the young grows. The greenish colouration of the Green Ringtail Possum is due to an unusual combination of black, white and yellow pigments and the structure of its hairs; and so differs from the Tree Sloth of South America whose renowned greenish fur is due to algal growth. As the young possum becomes more and more confident it plays around and about, but soon scampers back to the safety of its mother's body should danger threaten. Eventually the young reaches an age and size when it leaves its parents altogether and becomes quite independent.

It is a shock indeed to see kangaroos clambering through the leafy tree tops, but some members of the family have taken to such an unusual arboreal existence. In two small areas of tropical rainforest of the north live the Lumholtz's and Bennett's Tree Kangaroos. Unlike their ground dwelling counterparts they have stout, powerful and heavily clawed fore-legs for gripping branches, and a long thick tail to provide a counterbalance whilst jumping and climbing through the canopy in search of leafy foods. As well as hopping like terrestrial species, tree kangaroos can also move their hind legs independently. The continual clearing of our rainforest is sadly dimishing the already relatively small stands of rainforest occupied by these unique animals.

Green ringtail possum.

Red-legged pademelon.

POSTSCRIPT

Elsewhere on earth, tropical rainforest is almost exclusively found in third world countries where human population and economic pressures result in the irreversible clearing and degradation of rainforest for timber and land (which is poor beneath rainforest and of short-term agricultural value). Australia lacks these pressures, and is fast becoming the only politically and physically safe nation in the world to which people can come and enjoy tropical rainforests — the richest of all earths environments. In this respect we as a nation should recognise not only the potential recreational, aesthetic, educational and scientific benefits of rainforest but also its vast economic potential for recreational and educational tourism.

Having already destroyed at least half of Australia's rainforests it is surely possible and desirable to now conserve and manage what little remains as a natural, national and international heritage for all future generations to enjoy, appreciate, study and benefit from.

TAKE A WALK IN THE RAINFOREST
IN SOME QUEENSLAND NATIONAL PARKS
(listed from south to north)

SUBTROPICAL RAINFOREST PARKS
- Burleigh Heads
- Lamington, Cunningham's Gap and other Scenic Rim Parks
- Tambourine Mountains
- D'Aquilar
- Bunya Mountains
- Conondale
- Noosa
- Fraser Island

TROPICAL RAINFOREST PARKS
- Eungella
- Mt. Spec, at Paluma
- Hinchinbrook Island
- Dunk Island
- Wallaman Falls and other Herbert River Parks
- Clumpp Point
- Palmerston
- Ballenden Kerr
- Lake Barrine, Lake Eacham, and Hypipamee on Atherton Tableland
- Greater Daintree including Daintree River and Cape Tribulation
- Cedar Bay
- Iron Range
- Jardine River

Dragonfly

Stag beetle.

Red-eyed tree frog.

FURTHER READING

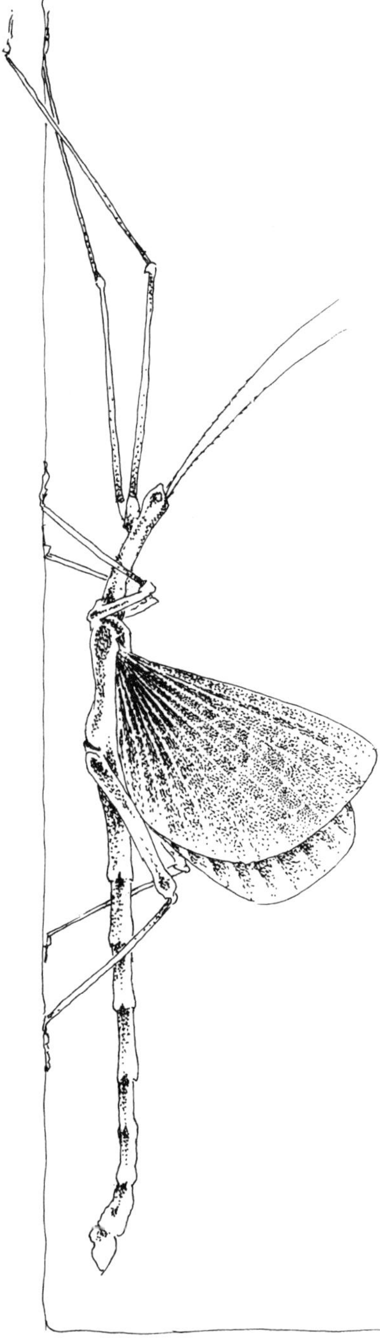

BREEDEN, S. & K. 1982
A Natural History of Australia: 1. Tropical Queensland
Second Printing, Collins, Sydney.

COGGER, H.G. 1983
Reptiles and Amphibia of Australia
Second Edition, Angus & Robertson, Sydney

FRITH, C.B. & D.W. 1983
Australian Tropical Rainforest Life
Speciality Education Supplies, Townsville

FRITH, C. & D. 1985
Australian Tropical Birds
Tropical Australia Graphics, Paluma

JONES, D.L. & CLEMESHA, S.C. 1981
Australian Ferns and Fern Allies
Revised Edition, Reed, Sydney.

PIZZEY, G. 1980
A Field Guide to the Birds of Australia
Collins, Sydney

READERS DIGEST SERVICES 1976
Readers Digest Complete Book of Australian Birds
Readers Digest, Sydney

RUSSELL, R. 1980
Spotlight on Possums
University of Queensland Press, Brisbane

RUSSELL, R. 1985
Daintree — Where Rainforest Meets Reef
Kevin Weldon & Australian Conservation Foundation, Sydney

STRAHAN, R. (Ed.) 1983
The Australian Museum Complete Book of Australian Mammals
Angus & Robertson, Sydney.

Stick insect.

Index of illustrations and major subjects

Eastern spinebill.

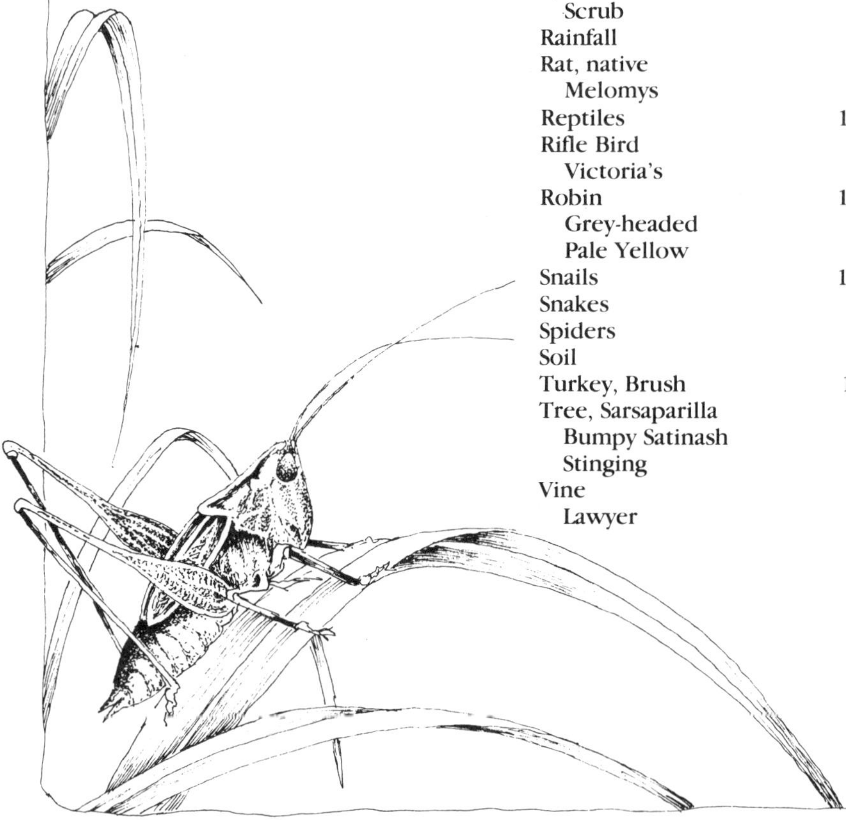

Grass Hopper